United States Government Accountability Office

Testimony
Before the Subcommittee on Communications and Technology, Committee on Energy and Commerce, House of Representatives

For Release on Delivery
Expected at 2:00 p.m. EDT
Tuesday, May 21, 2013

TELECOMMUNICATIONS NETWORKS

Addressing Potential Security Risks of Foreign-Manufactured Equipment

Statement of Mark L. Goldstein, Director
Physical Infrastructure Issues

GAO-13-652T

GAO Highlights

Highlights of GAO-13-652T, a testimony before the Subcommittee on Communications and Technology, Committee on Energy and Commerce, House of Representatives

May 21, 2013

TELECOMMUNICATIONS NETWORKS
Addressing Potential Security Risks of Foreign-Manufactured Equipment

Why GAO Did This Study

The United States is increasingly reliant on commercial communications networks for matters of national and economic security. These networks, which are primarily owned by the private sector, are highly dependent on equipment manufactured in foreign countries. Certain entities in the federal government view this dependence as an emerging threat that introduces risks to the networks. GAO was requested to review actions taken to respond to security risks from foreign-manufactured equipment.

This testimony addresses (1) how network providers and equipment manufacturers help ensure the security of foreign-manufactured equipment used in commercial communications networks, (2) how the federal government is addressing the risks of such equipment, and (3) other approaches for addressing these risks and issues related to these approaches.

This is a public version of a sensitive report that GAO issued in May 2013. Information deemed sensitive has been omitted. For the May 2013 report, GAO reviewed laws and regulations and interviewed officials from federal entities with a role in addressing cybersecurity or international trade, the five wireless and five wireline network providers with the highest revenue, and the eight manufacturers of routers and switches with the highest U.S. market shares. GAO obtained documentary and testimonial evidence from governmental entities in Australia, India, and the United Kingdom, because of their actions to protect their networks from supply chain attacks.

View GAO-13-652T. For more information, contact Mark Goldstein at (202) 512-2834 or goldsteinm@gao.gov.

What GAO Found

The network providers and equipment manufacturers GAO spoke with reported taking steps in their security plans and procurement processes to ensure the integrity of parts and equipment obtained from foreign sources. Although these companies do not consider foreign-manufactured equipment to be their most pressing security threat, their brand image and profitability depend on providing secure, reliable service. In the absence of industry or government standards on the use of this equipment, companies have adopted a range of voluntary risk-management practices. These practices span the life cycle of equipment and cover areas such as selecting vendors, establishing vendor security requirements, and testing and monitoring equipment. Equipment that is considered critical to the functioning of the network is likely to be subject to more stringent security requirements, according to these companies. In addition to these efforts, companies are collaborating on the development of industry security standards and best practices and participating in information-sharing efforts within industry and with the federal government.

The federal government has begun efforts to address the security of the supply chain for commercial networks. In 2013, the President issued an Executive Order to create a framework to reduce cyber risks to critical infrastructure. The National Institute of Standards and Technology (NIST)—a component within the Department of Commerce—is responsible for leading the development of the cybersecurity framework, which is to provide technology-neutral guidance to critical infrastructure owners and operators. NIST published a request for information in which NIST stated it is conducting a comprehensive review to obtain stakeholder input and develop the framework. NIST officials said the extent to which supply chain security of commercial communications networks will be incorporated into the framework is dependent in part on the input it receives from stakeholders. GAO identified other federal efforts that could impact communications supply chain security, but the results of those efforts were considered sensitive.

There are a variety of other approaches for addressing the potential risks posed by foreign-manufactured equipment in commercial communications networks, including those approaches taken by foreign governments. For example, the Australian government is considering a proposal to establish a risk-based regulatory framework that requires network providers to be able to demonstrate competent supervision and effective controls over their networks. The government would also have the authority to use enforcement measures to address noncompliance. In the United Kingdom, the government requires network and service providers to manage risks to network security and can impose financial penalties for serious security breaches. While these approaches are intended to improve supply chain security of communications networks, they may also create the potential for trade barriers, additional costs, and constraints on competition, which the federal government would have to take into account if it chose to pursue such approaches.

_____ United States Government Accountability Office

Chairman Walden, Ranking Member Eshoo, and Members of the Subcommittee:

Thank you for the opportunity to testify at today's hearing on federal and industry efforts related to communications supply chain security. The United States, like many other nations, is reliant on commercial communications networks for business and personal communication as well as for matters of national and economic security. Public and private organizations rely on computer systems to transmit sensitive and proprietary information, develop and maintain intellectual capital, conduct operations, process business transactions, transfer funds, and deliver services. In addition, the Internet has grown increasingly important to American business and consumers, serving as a medium for hundreds of billions of dollars of commerce each year. Many communications-based applications and services, including local and long-distance telephone calls, email, text messages, file transfers, and on-demand video programming, depend on effectively operating communications networks. Government, industry, and the public rely on communications networks to such a great degree that federal policy has included them in a category of national assets deemed critical infrastructure,[1] making their protection a national priority.[2] Many other critical infrastructure sectors such as banking and finance, energy, transportation systems, and water also rely

---

[1]The Uniting and Strengthening America by Providing Appropriate Tools Required to Intercept and Obstruct Terrorism Act of 2001, Pub. L. No. 107-56, § 1016(e), 115 Stat. 272, 401 (2001), codified at 42 U.S.C. § 5195c(e), defines critical infrastructure as the "systems and assets, whether physical or virtual, so vital to the United States that the incapacity or destruction of such systems and assets would have a debilitating impact on security, national economic security, national public health or safety, or any combination of those matters," which is incorporated by reference by section 2(4) of the Homeland Security Act of 2002, Pub. L. No. 107-296, § 2(4), 116 Stat 2135, 2140 (2002), codified at 6 U.S.C. § 101(4).

[2]The White House, *Presidential Decision Directive/NSC 63* (Washington, D.C.: May 1998). The White House, *Homeland Security Presidential Directive 7* (Washington, D.C.: December 2003).

on communications networks to sustain their operation.[3] In addition, we have identified protecting systems that support our nation's cyber critical infrastructure as a government-wide high-risk area.[4]

U.S. communications networks are, by and large, owned, operated, and safeguarded by the private sector. Network providers are dependent on a global supply chain[5] to provide equipment—such as routers, switches, and elements of evolved packet cores[6]—that is used to transport a high volume of aggregated voice and data traffic over their commercial communications networks. According to several network providers, very little of this equipment is manufactured in the United States. Equipment manufacturers—including those headquartered in the United States—are heavily dependent on facilities in foreign countries to design, manufacture, and assemble their products. This dependence on foreign-

---

[3]Federal policy established 18 critical infrastructure sectors: agriculture and food; banking and finance; chemical; commercial facilities; communications; critical manufacturing; dams; defense industrial base; emergency services; energy; government facilities; information technology; national monuments and icons; nuclear reactors, materials and waste; postal and shipping; public health and health care; transportation systems; and water. Homeland Security Presidential Directive 7 identified 17 critical infrastructure sectors, and the Department of Homeland Security (DHS) added critical manufacturing using authority provided under the directive. The White House, *Homeland Security Presidential Directive 7* (Washington, D.C.: December 2003) and Department of Homeland Security, *National Infrastructure Protection Plan: Partnering to enhance protection and resiliency* (2009).

[4]GAO's biennial high-risk list identifies government programs that have greater vulnerability to fraud, waste, abuse, and mismanagement or need transformation to address economy, efficiency, or effectiveness challenges. We have designated federal information security as a government-wide high-risk area since 1997; in 2003, we expanded this high-risk area to include protecting systems supporting our nation's critical infrastructure—referred to as cyber-critical infrastructure protection, or cyber CIP. See, most recently, GAO, *High-Risk Series: An Update*, GAO-13-283 (Washington, D.C.: February 2013).

[5]The National Institute of Standards and Technology (NIST) has defined the term "supply chain" to mean a linked set of resources and processes between acquirers, integrators, and suppliers that begins with the design of information and communications technology (ICT) products and services and extends through development, sourcing, manufacturing, handling, and delivery of ICT products and services to the acquirer. *Notional Supply Chain Risk Management Practices for Federal Information Systems (October 2012)* at http://nvlpubs.nist.gov/nistpubs/ir/2012/NIST.IR.7622.pdf.

[6]The evolved packet core is the core network used for long-term evolution (LTE) systems; a standard for commercial wireless technologies. LTE is widely accepted as the foundation for future mobile communications.

manufactured equipment[7] is viewed by some federal entities as an emerging threat that introduces potential risks[8] to the networks.[9] According to the Office of the National Counterintelligence Executive, "the globalization of the economy has placed critical links in the manufacturing supply chain under the direct control of U.S. adversaries."[10] A potential enemy or criminal group has a number of ways to potentially exploit vulnerabilities in the communications equipment supply chain, such as placing malicious code in the components that could compromise the security and resilience of the networks.[11]

Recent government efforts in the United States and other countries highlight concerns about the potential impact of supply chain threats on government, industry, and personal communications and transactions. Legislative proposals in the United States have sought to improve the protection of critical infrastructure, such as commercial communications

---

[7]For the purpose of this report, we define foreign-manufactured equipment as equipment produced, either in whole or in part, outside of the United States.

[8]NIST defines "threat" as any circumstance or event with the potential to adversely affect organizational operations and assets, individuals, other organizations, or the nation through an information system via unauthorized access, destruction, disclosure, modification of information, or denial or disruption of service. According to NIST, risk is a measure of the extent to which an entity is threatened by a potential circumstance or event, and typically a function of (1) the adverse impacts that would arise if the circumstance or event occurs, and (2) the likelihood of occurrence, which is based on an analysis of the probability that a given threat is capable of exploiting a given vulnerability. NIST also defines "vulnerability" as a weakness in an information system, system security procedures, internal controls, or implementation that could be exploited by a threat. Department of Commerce, National Institute of Standards and Technology, *Glossary of Key Information Security Terms* (Washington D.C.: 2011).

[9]White House Cyberspace Policy Review, *Assuring a Trusted and Resilient Information and Communications Structure*. http://www.whitehouse.gov/assets/documents/Cyberspace_Policy_Review_final.pdf.

[10]Office of the National Counterintelligence Executive, *Supply Chain Threats*, accessed on January 28, 2013, http://www.ncix.gov/issues/supplychain/index.php.

[11]Supply chain-related threats can be introduced in the manufacturing, assembly, and distribution of hardware, software, and services. We are not addressing disruptions that can be caused by labor or political disputes and natural causes (e.g., earthquakes, fires, floods, or hurricanes) that could affect the availability of equipment that is used to support the communication networks.

networks, from cyber attacks.[12],[13] Likewise, the White House released an Executive Order and a presidential policy directive in February 2013 that seek to improve the protection of critical infrastructure, including communications networks, from cyber attacks.[14] In 2012, the House Committee on Energy and Commerce, Subcommittees on Oversight and Investigations, and Communications and Technology held a series of hearings that addressed, among other things, cybersecurity[15] threats to communication networks.[16] The House Permanent Select Committee on Intelligence released a report in October 2012 in which it recommended the United States view with suspicion the continued penetration of the U.S. telecommunications market by Chinese telecommunications companies.[17] To help protect against the potential national security risks, the committee further recommended that U.S.-based network providers

---

[12]NIST defines "cyber attack" as an attack, via cyberspace, targeting an enterprise's use of cyberspace for the purpose of disrupting, disabling, destroying, or maliciously controlling a computing environment/infrastructure, or destroying the integrity of the data or stealing controlled information.

[13]Cyber Intelligence Sharing and Protection Act, H.R. 624, 113th Cong. (2013); the Cybersecurity and American Cyber Competitiveness Act, S. 21, 113th Cong. (2013).

[14]*Improving Critical Infrastructure Cybersecurity*, Exec. Order No. 13,636, 78 Fed. Reg. 11,739 (February 12, 2013). *Directive on Critical Infrastructure Security and Resilience*, Presidential Policy Directive 21, 2013 Daily Comp. Pres. Docs. No. 92. (February 12, 2013).

[15]According to NIST, "cybersecurity" means the ability to protect or defend the use of "cyberspace" from cyber attacks. NIST defines "cyberspace" as a global domain within the information environment consisting of the interdependent network of information systems infrastructures including the Internet, telecommunications networks, computer systems, and embedded processors and controllers.

[16]House Committee on Energy and Commerce, Subcommittee on Oversight and Investigations, hearing on *IT Supply Chain Security: Review of Government and Industry Efforts* (Mar. 27, 2012). House Committee on Energy and Commerce, Subcommittee on Communications and Technology, hearings on *Cybersecurity and the Pivotal Role of Communications Networks*, March 7, 2012; and *Cybersecurity: Threats to Communications Networks and Public-Sector Responses*, March 28, 2012.

[17]Permanent Select Committee on Intelligence, U.S. House of Representatives, *Investigative Report on the U.S. National Security Issues Posed by Chinese Telecommunications Companies Huawei and ZTE*, (Washington, D.C.: Oct. 8, 2012). The report states that the Chinese government or intelligence services could access equipment during the production process to insert malicious hardware or software for economic or foreign espionage with or without the cooperation of the companies. The report contains a classified annex that provides more information regarding the Committee's concerns about the risk. We did not access the annex.

consider the long-term security risks associated with purchasing products or services from specific foreign-based equipment manufacturers. Other countries—such as Australia, India, and the United Kingdom—are similarly concerned about the emerging threats to their commercial communication networks posed by the global supply chain and have taken actions to improve their ability to address this security challenge.

You asked us to examine private-sector and government actions to respond to the potential security risks posed by the use of foreign-manufactured equipment. This testimony is a public version of a sensitive report that we issued in May 2013 in response to your request. This testimony communicates the publicly releasable aspects of our findings while omitting information considered sensitive regarding federal actions taken to address potential security risks from foreign-manufactured equipment. This testimony discusses the objectives of our report, which were to examine:

> 1) How communications network providers and equipment manufacturers help ensure the security of foreign-manufactured equipment used in commercial communications networks.
>
> 2) How the federal government is addressing the potential risks of the use of foreign-manufactured equipment in commercial communications networks.
>
> 3) Other approaches for addressing the potential risks of the use of foreign-manufactured equipment in commercial communications networks and issues related to these approaches.

In preparing this statement, we relied on the work supporting our May 2013 report. For that report, we interviewed and collected documentation from federal agencies, including the Department of Commerce (Commerce), the Department of Homeland Security (DHS), the Department of Defense (DOD), the General Services Administration (GSA), and the Federal Communications Commission (FCC), among others, that have a role in addressing cybersecurity to identify federal efforts to address the risks of using foreign-manufactured equipment in commercial communications networks. We also asked federal agencies to identify statutes and regulations related to the federal government's legal and regulatory authority over how communications network providers ensure the security of their U.S. commercial networks.[18] We interviewed commercial communications network providers and equipment manufacturers that supply providers with routers, switches, and evolved packet cores to discuss their approaches for ensuring the security of the equipment used in commercial communications networks. We focused this work on the five wireless and five wireline network providers with the highest revenue and the eight manufacturers of routers and switches with the largest market share[19] in the United States. We did not test the effectiveness of the practices identified by the federal government, communication network providers, or equipment manufacturers.

Additionally, through a review of government and academic studies and interviews with stakeholders, we identified and described other approaches from governmental entities in Australia, India, and the United Kingdom that address supply chain risks for commercial communications networks.[20] We chose these countries to show the variation in how foreign governments are approaching supply chain risk management and because of the availability of public information in English describing their approaches. While the results of the data collected from these three countries may not encompass all possible approaches, they provided important insights into the approaches that some countries are using to address supply chain risks for commercial communications networks. We also assessed the potential for using the Committee on Foreign Investment in the United States (CFIUS)[21] review process for purchases of foreign-manufactured equipment. A voluntary notification process similar to CFIUS is being discussed by government and industry

---

[18]This report focuses on the wireline, wireless, and cable networks, and the core routing and switching equipment within those networks because they represent the majority of traffic.

[19]The eight manufacturers of routers and switches had a combined market share of 92 percent. We did not have access to data on market share for wireline and wireless providers.

[20]We attempted to include Canada in our review, but there was limited public information on its approach, and Canadian officials did not respond to our request for an interview.

[21]CFIUS is an inter-agency committee, established by Exec. Order No. 11858, 40 Fed. Reg. 20,263 (1975), as amended, authorized to review transactions that could result in control of a U.S. business by a foreign person, in order to determine the effect of such transactions on the national security of the United States.

stakeholders. We reviewed the Foreign Investment and National Security Act of 2007,[22] related regulations, and CFIUS annual reports to Congress to describe the CFIUS process and its applicability to purchases of foreign equipment for commercial networks. Finally, we conducted our own analysis regarding several potential issues that could arise from the use of these approaches. We identified these issues based on interviews with foreign government officials and U.S. industry stakeholders, and our review of foreign proposals and other documentation. The issues identified do not present an exhaustive list of all issues that could arise, but rather provide a range of considerations involved in other approaches to addressing supply chain risks.

We conducted this work from December 2011 to May 2013 in accordance with generally accepted government auditing standards. Those standards require that we plan and perform the audit to obtain sufficient, appropriate evidence to provide a reasonable basis for our findings and conclusions based on our audit objectives. We believe that the evidence obtained provides a reasonable basis for our findings and conclusions based on our audit objectives. See appendix I for more information about our scope and methodology.

Background

Cybersecurity and Critical Infrastructure Protection Responsibilities

Federal policy calls for critical infrastructure protection activities that are intended to enhance the cyber and physical security of private infrastructures, such as telecommunication networks, that are essential to national and economic security. DHS, Commerce, and FCC have critical infrastructure protection responsibilities over issues related to the security of communications networks.[23] Appendix IV provides additional

---

[22] Pub. L. No. 110–49, 121 Stat. 246 (2007). *See, also,* 50 App. U.S.C. § 2061 note.

[23]Federal agencies can impose conditions on companies with which they contract. Network service providers and equipment manufacturers therefore may be subject to security requirements that are specific to contracts they have with the federal government. GSA officials told us that the Office of Management and Budget requires GSA to include supply-chain risk-management language in some of its critical-infrastructure-related contracts. The language requires documentation of a product's manufacturing chain of custody. However, according to GSA officials, this language is limited to critical-infrastructure-related contracts because of the higher cost of meeting the requirements.

information on these agencies' legal authority related to supply chain security for commercial communication networks. In addition, some executive actions have focused on supply chain risk management issues related to cybersecurity, which are described below.

Department of Homeland Security

The Homeland Security Act of 2002[24] established DHS and assigned it the following critical infrastructure protection responsibilities:

- develop a comprehensive national plan for securing the key resources and critical infrastructure of the United States and
- disseminate, as appropriate, information to assist in the deterrence, prevention, and pre-emption of or response to terrorist attacks.[25]

Department of Commerce

Commerce is responsible under Presidential Policy Directive 21 (PPD-21), in coordination with other federal and nonfederal entities, for improving security for technology and tools related to cyber-based systems, and promoting the development of other efforts related to critical infrastructure to enable the timely availability of industrial products, materials, and services to meet homeland security requirements.[26] Within Commerce, the National Institute of Standards and Technology (NIST) has responsibility for, among other things, cooperating with other federal agencies, industry, and other private organizations in establishing standard practices, codes, specifications, and voluntary consensus standards.[27]

Federal Communications Commission

Under PPD-21, FCC is responsible for exercising its authority and expertise to partner with other federal agencies on:

- identifying and prioritizing communications infrastructure;
- identifying communications sector vulnerabilities and working with industry and other stakeholders to address those vulnerabilities; and

---

[24] 6 U.S.C. ch. 1.

[25] Homeland Security Act, § 201, 6 U.S.C. § 121(d)(5), (8).

[26] The White House, *Presidential Policy Directive 21* (Washington, D.C.: February 2013). Prior to PPD-21, Commerce was responsible under Homeland Security Presidential Directive 7, in coordination with other federal and nonfederal entities, for improving technology for cyber systems and promoting critical infrastructure efforts. The White House, Homeland Security Presidential Directive 7 (Washington, D.C.: December 2003).

[27] 15 U.S.C § 272.

	- working with stakeholders, including industry, and engaging foreign governments and international organizations to increase the security and resilience of critical infrastructure within the communications sector and facilitating the development and implementation of best practices promoting the security and resilience of the nation's critical communications infrastructure.[28]
Executive Actions	Supply chain risk management has been the focus of executive actions; for example, in January 2008, the President directed the development of a multi-pronged approach for addressing global supply chain risk management as part of the Comprehensive National Cybersecurity Initiative (CNCI), an ongoing effort.[29] More recently, at the direction of the President, a report on the federal government's cybersecurity-related activities was released, which discussed, among other things, the importance of prevention and response against threats to the supply chains used to build and maintain the nation's infrastructure.[30] Additionally, in response to one of the report's recommendations, the President appointed a national cybersecurity coordinator in December 2009.
Description of Core Networks and Access Networks	The United States has several nationwide voice and data networks that along with comparable communications networks in other countries, enable people around the world to connect to each other, access information instantly, and communicate from remote areas. These networks consist of core networks,[31] which transport a high volume of aggregated voice and data traffic over significant distances, and access networks, which are more localized and connect end users to the core network or directly to each other. Multiple network providers in the United

[28]The White House, *Presidential Policy Directive 21* (Washington, D.C.: February 2013).

[29]The White House, *National Security Presidential Directive 54/Homeland Security Presidential Directive 23*. (Washington, D.C.: Jan. 8, 2008).

[30]The White House, *Cyberspace Policy Review: Assuring a Trusted and Resilient Information and Communications Infrastructure*. May 2009. http://www.whitehouse.gov/assets/documents/Cyberspace_Policy_Review_final.pdf.

[31]NIST officials stated that there are no agreed-upon definitions of "core network," "core equipment," or "core infrastructure." The descriptions of the terms in this report are based on information in the *2012 Risk Assessment Report for Communications*, published by DHS's National Communications System.

States operate distinct core and access networks that interconnect to form a national communications infrastructure (see fig. 1).

Figure 1: Communications Core and Access Networks

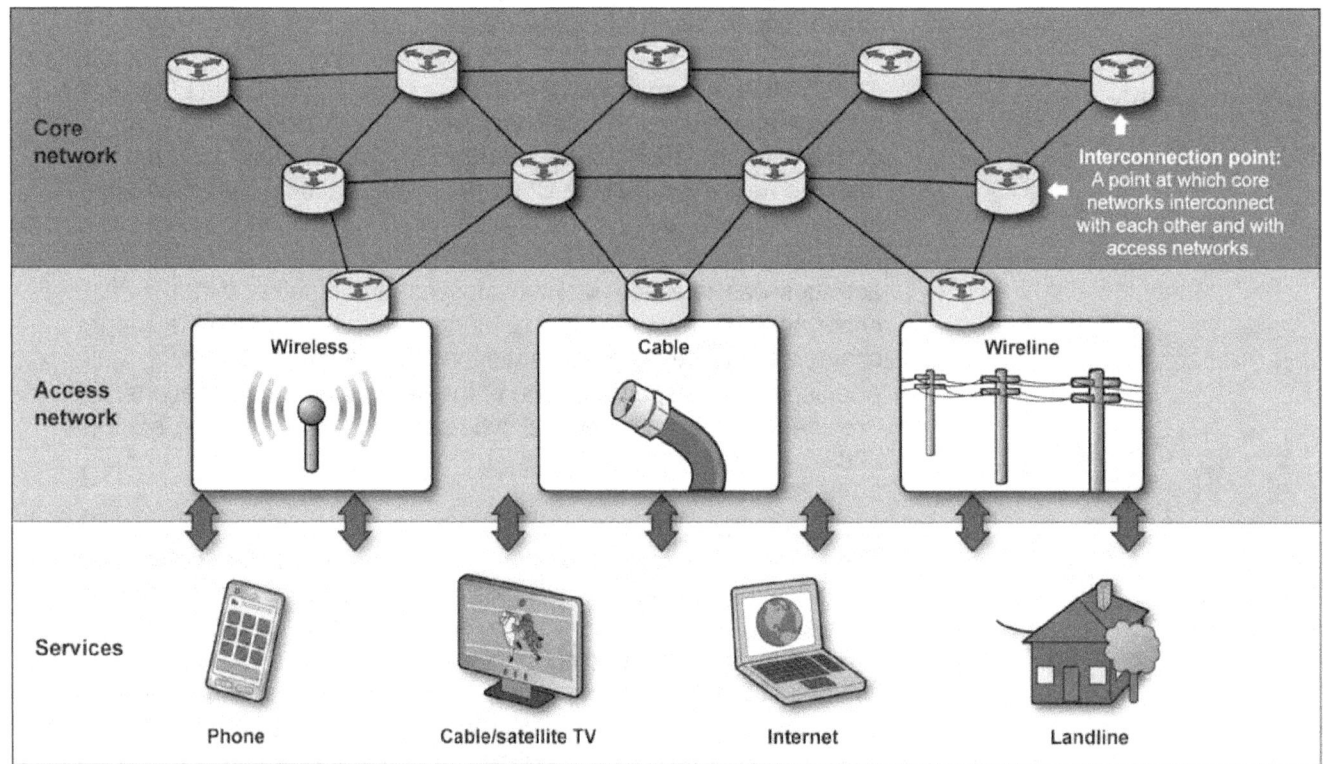

Sources: DHS and GAO.

Note: As stated previously, this testimony discusses only the wireline, wireless, and cable access segments of the communications sector.

Routers and switches send traffic, in the form of data packets, through core and access networks. These pieces of equipment read the address information located in the data packet, determine its destination, and direct it through the network. Routers connect users between networks, while switches connect users within a network.[32] The evolved packet core is the mobile core network used for long-term evolution (LTE) systems, a

---

[32]Many switches are now designed to perform the functions of routers as well as other security services such as firewalls and intrusion detection.

	standard for commercial wireless technologies. LTE is widely accepted as the foundation for future mobile communications. Several major network equipment manufacturers are competing to provide equipment to wireless network providers that are upgrading their networks to deploy LTE.
Global Supply Chain	Communications infrastructure is increasingly composed of components that are designed, developed, and manufactured by foreign companies or by U.S. companies that rely on suppliers that integrate foreign components into their products.[33] Furthermore, we have previously reported that according to NIST, today's complex global economy and manufacturing practices make corporate ownership and control more ambiguous when assessing supply chain vulnerabilities, as companies may conduct business under different names in multiple countries.[34] For example, foreign-based companies sometimes manufacture and assemble products and components in the United States, and U.S.-based companies sometimes manufacture products and components overseas or employ foreign workers domestically. Figure 2 depicts some of the locations that major network equipment manufacturers we spoke with use for different steps in the production process.

[33]Telcordia, *Mitigating the Supply Chain Security Risks in National Public Telecommunications Infrastructure,* (2011).

[34]GAO, *IT Supply Chain: National Security-Related Agencies Need to Better Address Risks*, GAO-12-361 (Washington, D.C.: March 23, 2012).

Figure 2: Examples of Supply Chain Locations for Network Equipment Manufacturers

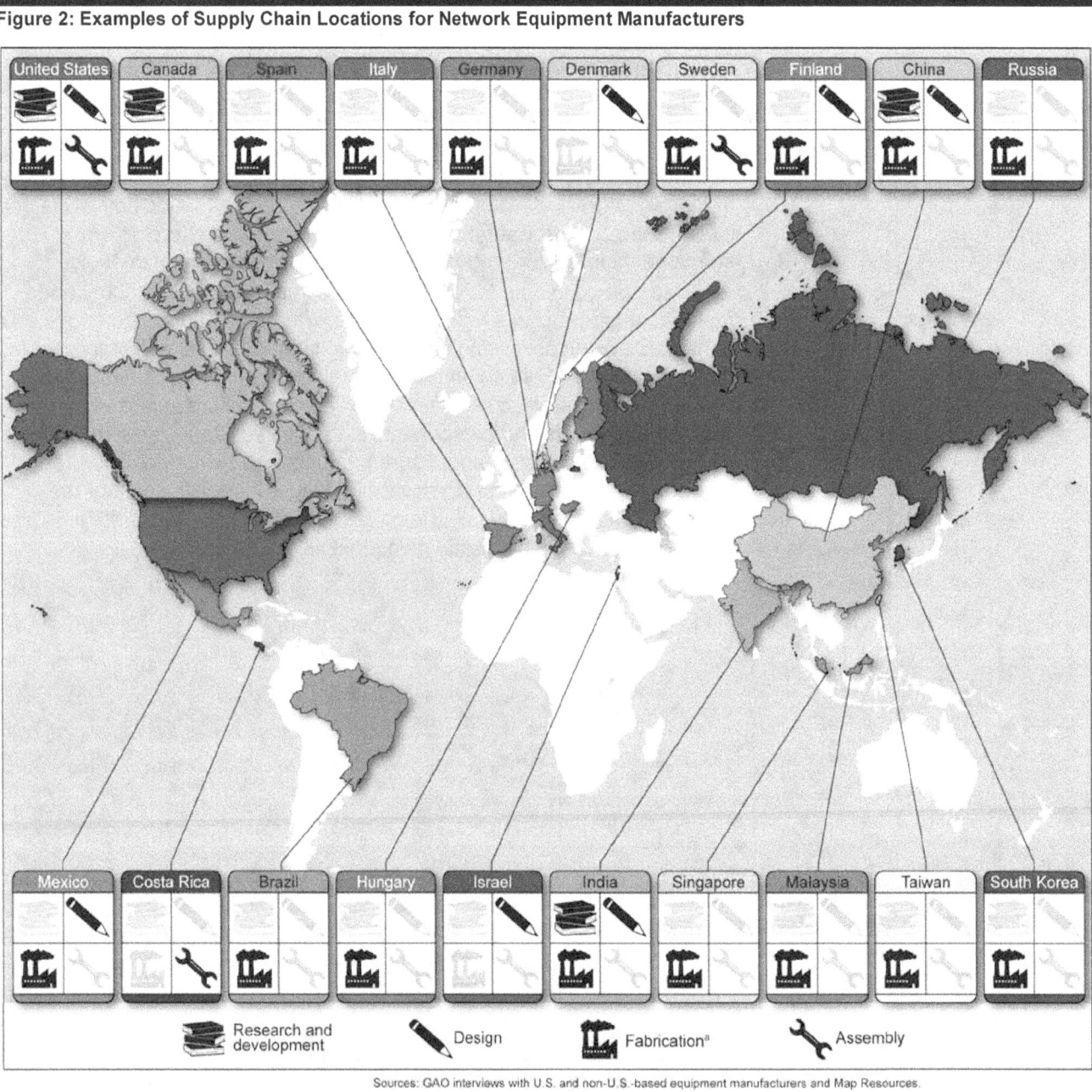

Note: Bold icons indicate that the production step is conducted in the country.

[a] Fabrication is the construction of a physical item from raw materials or the lowest-level parts.

From 2007 through 2011, communications network equipment imported for the U.S. market came from over 100 foreign countries.[35] While the import data do not distinguish whether the imports are from U.S. or foreign-based companies, according to International Trade Commission staff, many of the imports are from U.S. companies manufacturing abroad. Imports of network equipment to the United States grew about $10 billion (about 76 percent) over a 5-year period, from $13.5 billion in 2007 to $23.8 billion in 2011, as shown in figure 3. During this same period, imports from China, which was the leading source country, grew by $4.9 billion (112 percent). In 2011, the top five sources of U.S. imports of networking equipment were China ($9.3 billion), Mexico ($5.2 billion), Malaysia ($2.6 billion), Thailand ($1.9 billion), and Canada ($713 million).

---

[35]U.S. International Trade Commission. *Interactive Tariff and Trade DataWeb* (accessed Dec. 5, 2012). [Data file]. http://dataweb.usitc.gov/. The data are based upon a search for Harmonized Tariff Schedule (HTS) Code 851762, which includes machines for the reception, conversion, and transmission or regeneration of voice, images, or other data, including switching and routing apparatus.

Figure 3: Total U.S. Imports of Network Equipment and Top Five Sources by Country, 2007 to 2011

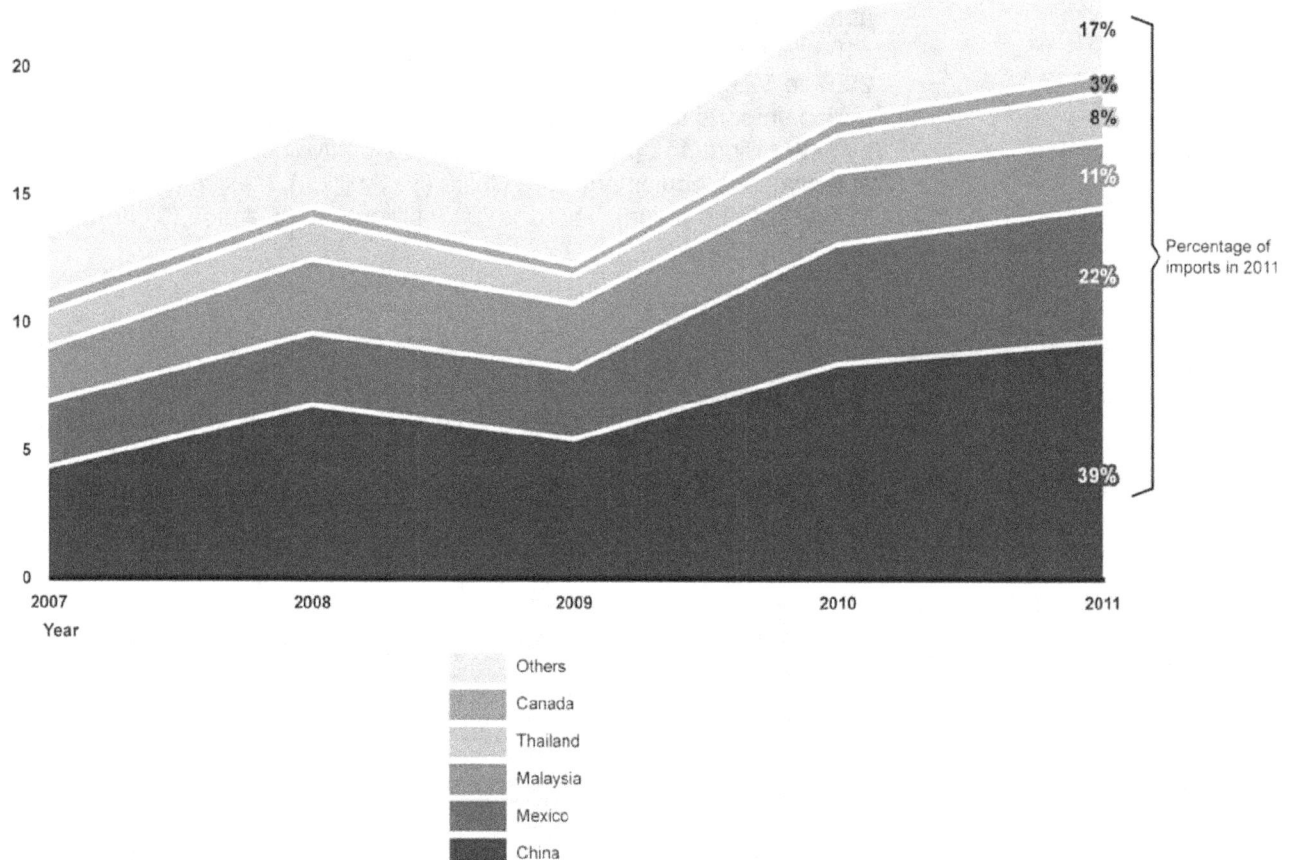

Source: GAO analysis of U.S. International Trade Commission staff data.

Note: The data included imports that were characterized as Harmonized Tariff Schedule Code 851762, which includes machines for the reception, conversion, and transmission or regeneration of voice, images, or other data, including switching and routing apparatus.

While there is no comprehensive unclassified compilation of attacks to core networks that originated in the supply chain,[36] reliance on a global

---

[36] Network providers may be reluctant to publicly divulge this information because of business concerns. For those incidents publicly reported, it can be difficult to discern if the attack was targeted to core network equipment.

supply chain introduces some degree of risk. Risks include threats posed by actors such as foreign intelligence services or counterfeiters that may exploit vulnerabilities in the supply chain, thus compromising the availability, security, and resilience of the networks.[37] Multiple points in the supply chain may present vulnerabilities that threat actors could exploit. For example, a lack of adequate testing for software patches and updates could leave a communications network vulnerable to the insertion of code intended to allow unauthorized access to information on the network. Routers and switches can present points of vulnerability because they connect to the core network and are used to aggregate data, according to an FCC official with whom we spoke. For example if a threat actor gained control of a router, that actor could disrupt data traffic to and inside core networks. Supply chain threats and vulnerabilities are discussed in more depth in appendixes II and III, respectively.

Industry Is Addressing the Risks of Using Foreign-Manufactured Equipment

Companies Address Supply Chain Risk through Procurement and Testing Practices

The network providers and equipment manufacturers we met with told us they address the potential security risks of using foreign-manufactured equipment through voluntary risk management practices. Officials from the companies and industry groups that we spoke with said that they consider the level of risk to be affected not by where equipment and components are made, but how they are made, particularly the security procedures implemented by manufacturers. Many of these officials also said they were not aware of any intentional attacks originating in the supply chain, and some said that they consider the risk of this type of attack to be low. Officials from four industry groups and one research institution we spoke with told us that supply chain attacks are harder to carry out and require more resources than other modes of attacks, such

---

[37]Supply chain-related threat actors include corporate spies, corrupt government officials, cyber vandals, disgruntled employees, foreign military, government agents or spies, radical activists, purveyors of counterfeit goods, or criminals. GAO-12-361.

as malicious software uploaded to equipment through the Internet, and, therefore, are the less likely vehicle to be used by potential attackers.[38] Three network providers told us the most common anomalies found in equipment are caused by erroneous coding in the software, anomalies that are unintentional. Such anomalies could, however, lead to exploitable vulnerabilities, according to officials from a third-party testing firm.[39] Nonetheless, the companies we spoke with told us that security is a high priority because their brand image and profitability depends, in part, on avoiding any type of breach of security or disruption of service.

Network providers and equipment manufacturers told us that their voluntary risk management practices are in the areas of vendor selection, vendor security requirements, and equipment testing and monitoring, as described below and in figure 4. They said these practices are often a part of their company's overall security plans and procurement processes and are applied throughout the entire life cycle of their equipment.[40]

---

[38]Officials from an industry group and a research institution, as well as a recent congressional report also noted that a likely threat actor to carry out a supply chain attack would be a nation-state, because it may have the capabilities and the incentives for conducting such attacks.

[39]According to a recent congressional report and an official from a research institution that we spoke with, sophisticated implants in equipment, such as inserting malicious code into firmware, along the supply chain may be very difficult to detect. Permanent Select Committee on Intelligence, U.S. House of Representatives, *Investigative Report on the U.S. National Security Issues Posed by Chinese Telecommunications Companies Huawei and ZTE*, (Washington, D.C.: Oct. 8, 2012).

[40]We did not test the effectiveness of these practices and have not described all the supply-chain risk-management practices that network providers and equipment manufacturers implement. Because we collected this information from the network providers and equipment manufacturers with the largest market shares, it may not be representative of the approaches taken by all companies.

Figure 4: Examples of Companies' Supply Chain Risk-Management Practices

Practice	Description of practice
Vendor selection	Companies carefully select vendors to ensure the security and reliability of their equipment using a number of considerations, including: • vendor's security practices • vendor's use of security-related standards • vendor's security reputation • criticality of equipment or services being procured
Vendor security requirements	Companies may require vendors to follow certain supply chain security practices that are based on perceived security risks. Examples include: • processes related to physical security of products • restrictions on vendor access to sensitive company information • employee verification practices Some companies also require the right to conduct inspections of a vendor's manufacturing sites and compliance with requirements.
Equipment testing and monitoring	Companies test equipment throughout its life cycle to detect and mitigate vulnerabilities. After equipment is implemented into the network, companies monitor traffic and equipment performance to detect any abnormal activity that may indicate a cyber attack or vulnerability.

Source: GAO.

Vendor Selection

The network providers and equipment manufacturers we spoke with said that ensuring the security and reliability of their equipment requires them to carefully select their vendors.[41] In addition to the typical considerations when selecting vendors—prices and product performance, the vendor's financial stability, and maintenance and service options offered—the providers and manufacturers told us that they consider security-related factors, such as the vendor's security practices, the industry standards related to security the vendors follow, and past security performance or reputation.[42] Another consideration for some network providers when selecting vendors is how critical the equipment being procured is to network operations. Components that will be used in the core network, for example, are typically purchased from vendors that network providers

---

[41] We refer to vendors in this section as those companies that supply network service providers with equipment or those that supply parts to equipment manufacturers.

[42] Network service providers and equipment manufacturers told us that there are quality-control and security-related industry standards for vendors that are not specific to supply chain, but do affect security, and a vendor's compliance with these maybe favorably viewed.

consider most trustworthy. Some network providers told us they also value having long-term relationships with equipment manufacturers, because they are able to develop trust over time that the manufacturer will provide them with reliable and secure equipment and services.

While network providers said that they are aware of security concerns about vendors from certain countries, they do not exclude vendors from consideration that have manufacturing locations in those countries, in part, because the global nature of the supply chain would make excluding all vendors located in a particular country difficult. Some network providers told us they may exclude or avoid vendors based on factors such as the ownership of the company or concerns about the security of the vendor's product, and two told us that federal government officials had advised against using specific vendors for national security reasons, as discussed in the following section of this testimony.

Vendor Security Requirements

Network providers and equipment manufacturers told us that once vendor selections are made, they might require vendors to follow certain security practices, often as part of their contracts. Network providers told us that the security practices they require are typically based on the criticality or perceived risk of the project and the role of the vendor. For example, one network provider we spoke with generates a vendor risk profile for purchases that it considers critical or high risk or if it does not have an established relationship with the vendor. The company uses the profile to collect information on the product or service being provided, the vendor's access to proprietary information, such as the company's financial information or customer sensitive information, and available information on a vendor's subcontractors. This information enables the network provider to identify areas of concern to investigate and to customize the security requirements placed on the vendor. The security practices that both network providers and equipment manufacturers may require of their vendors include the following:

- physical security measures, such as procedures for securing manufacturing sites, transporting equipment and parts, and packaging equipment and parts;
- access controls, such as limiting in-house and vendor employees' access to equipment, maintaining records of who accesses equipment, and restricting who performs patches and updates; and
- employee security measures, such as requiring employees to have background checks and use passwords and user verification to access systems.

Additionally, network providers and equipment manufacturers told us they might require vendors to allow inspections of their manufacturing sites to check for compliance with the agreed-upon security practices. Representatives from the companies we met with told us that they conduct inspections at varying frequencies and for a number of reasons, including if the vendor is providing a critical piece of equipment or part or is identified as high risk, or if the equipment is performing poorly.

Equipment Testing and Monitoring

Network providers and equipment manufacturers told us that equipment is tested to detect vulnerabilities. This is done throughout the life cycle of equipment, including during product development, before and after implementation, and when any patches or updates are applied. After equipment is installed into the network, network providers also monitor the equipment constantly to detect abnormal traffic or problems with the equipment that might indicate a potential cyber attack and disrupt network service. According to officials from a third-party testing firm, there are several tools available to test the security of equipment, including:

- *vulnerability scans*—searching software and hardware for known vulnerabilities;
- *penetration testing*—executing deliberate attempts to attack a network through the equipment, sometimes targeting specific vulnerabilities of concern; and
- *source code analysis*—evaluating in depth the underlying software code that can uncover unknown vulnerabilities that would not be detected during a vulnerability scan.[43]

Testing can be performed by the network provider, the equipment manufacturer, or independent third-party testing firms. Most network providers and several equipment manufacturers told us they use third-party testing firms on an ad-hoc basis, such as when requested by a customer or when they do not have the expertise or resources to conduct appropriate tests. Network providers and equipment manufacturers also use these firms when they want to analyze software or firmware source code because equipment manufacturers are reluctant to provide network

---

[43]There are other specialized tools available for certain situations. For example, officials from a third-party security firm told us that a network provider may conduct forensic analysis following a compromise of their network to provide a high level of assurance that the issue has been resolved.

providers with source code, which they consider intellectual property.[44] Two network providers and one equipment manufacturer told us they use a trusted delivery model that employs a third-party testing firm to ensure that the equipment purchased and received is secure. Under this model, the third-party testing firm tests equipment over the full life-cycle of equipment, including when there are software patches or hardware updates, and uses a number of different techniques, such as source code analysis. Additionally, the testing firm verifies that the equipment delivered and implemented by the network provider matches the equipment tested and that the equipment manufacturer followed certain security procedures.

However, a recent congressional report identified the following potential limitations of third-party testing and available testing techniques.

- These firms typically test equipment that is configured in a specific and restrictive way that may differ from the configuration that is actually deployed in the network.

- The behavior of equipment can vary widely depending on how and where it is configured, installed, and maintained.

- The pace of technology is changing more rapidly than third-party evaluation processes.

- Vendors that finance their own security evaluations create a conflict of interest that can lead to skepticism about the independence and rigor of the result.[45]

Officials from a third-party testing firm told us that there are evaluation processes, such as the trusted delivery model, that test the equipment delivered to network providers and deployed into the network against the equipment tested. Although they said it is impossible to test every piece of equipment, the firm tests a statistically significant random selection of equipment that represents all manufacturing lots and geographic

---

[44]Firmware is the combination of a hardware device and computer instructions and data that reside as read-only software on that device.

[45]Permanent Select Committee on Intelligence, U.S. House of Representatives, *Investigative Report on the U.S. National Security Issues Posed by Chinese Telecommunications Companies Huawei and ZTE*, (Washington, D.C.: Oct. 8, 2012).

locations. They also told us that independence is critical to their business. The officials said the vendor has no visibility into the evaluation process, and, typically, the vendor is obligated to report testing results.

The congressional report further stated that regardless of the testing technique employed, fully preventing a determined and clever insider from intentionally inserting flaws into equipment means finding and eliminating every significant vulnerability from a complex product, a monumental, or even—in the words of one congressional report—"virtually impossible" task.[46] Similarly, officials from one third-party testing firm whom we spoke with told us that they have concerns about the effectiveness of network monitoring as a way of detecting vulnerabilities. They said that security monitoring, in most cases, can only detect attempts to exploit known vulnerabilities, or in more complex approaches, identify potentially dangerous anomalous network activity. And as systems evolve and are updated, new vulnerabilities that have long existed in the underlying equipment may be inadvertently exposed in a manner that makes exploitation possible.

Companies Collaborate on Supply-Chain Security Standards, Best Practices, and Information Sharing

There are currently no industry standards that address all aspects of supply chain risk management, including supply chain security, and few best practices that provide industry with guidance on determining what practices to use. However, according to officials from companies and industry groups and the experts we spoke with, there are several industry-led efforts to establish standards and best practices and share information related to supply chain security.[47] Some network providers and equipment manufacturers told us that they developed their own practices based on national and international standards that address information systems' security, such as those practices described within

---

[46]Permanent Select Committee on Intelligence, U.S. House of Representatives, *Investigative Report on the U.S. National Security Issues Posed by Chinese Telecommunications Companies Huawei and ZTE*, (Washington, D.C.: Oct. 8, 2012).

[47]Stakeholders we spoke with told us about efforts related to securing the software supply chain, such as those conducted by the Software Assurance Forum for Excellence in Code, which is an industry-led group that develops best practices for reliable software, hardware, and services and DHS's Software Assurance Program. These groups have published several supply-chain security guidelines for the development of secure software.

the certification program called the Common Criteria,[48] and those developed by the International Organization for Standardization (ISO), the International Electrotechnical Commission (IEC), NIST, and the Internet Engineering Task Force. However, these standards are not specific to supply chain security.[49] Additionally, federal agencies that we have identified as having jurisdiction over issues related to the security of communications networks have not established supply chain security requirements for the communications industry, as discussed further in the next section of this testimony.[50] The companies we spoke with also told us they have been participating in information sharing about cybersecurity issues, including supply chain security, in venues including informal conversations, industry group meetings, and discussions with the federal

---

[48]The Common Criteria provides a common set of requirements for the security functionality of information technology (IT) products and for assurance measures applied to these IT products during evaluation. Evaluations of IT products are conducted by independent and licensed laboratories, and those that meet the Common Criteria requirements are provided with a certification. These certifications are recognized by participating member countries.

[49] According to a DOD official, there are a number of national and global standards-development organizations—such as ISO, the Common Criteria's technical working group, and the Common Criteria Development Board—that have supply-chain risk-management-related initiatives. According to officials from NIST and DOD, one of the more significant standards being developed is ISO/IEC 27036 "IT Security—Security techniques—Information security for supplier relationships." This draft standard will offer guidance on the evaluation and mitigation of security risks involved in the procurement and use of information or IT-related services supplied by other organizations. NIST officials told us that the proposed standard would address the risk management aspects of the entire ICT supply chain from the perspectives of suppliers and customers. DOD officials told us that all of the supply-chain risk- management initiatives and standards development activities are monitored and harmonized where possible.

[50]In October 2012, NIST published an interagency report that describes a set of supply-chain assurance methods and practices to help federal departments and agencies manage the associated information and communications technology (ICT) supply-chain risks over the entire life cycle of ICT systems, products, and services. NIST officials told us that they are developing a special publication related to this report. Several network providers and equipment manufacturers we spoke with said that these could serve as a reference for private companies to use when developing their own supply-chain risk-management practices. *Notional Supply Chain Risk Management Practices for Federal Information Systems* (October 2012) at http://nvlpubs.nist.gov/nistpubs/ir/2012/NIST.IR.7622.pdf.

government. Below are the two industry-led efforts most frequently discussed during our interviews.[51]

The Open Group Trusted Technology Forum (OTTF)

The OTTF is a forum within The Open Group, which is a global consortium that represents all sectors of the IT community including academics, equipment manufacturers, federal agencies, and software developers. The Open Group establishes certification programs and voluntary consensus standards, such as standards for security, enterprise architecture, interoperability, and systems management.[52] The OTTF's objective is to create and adopt standards to improve the security and integrity of commercial off-the-shelf information and communication products, including hardware and software, as they are being developed and moved through the global supply chain. In April 2013, the OTTF published a voluntary standard[53] that is intended to enhance the security of global supply chains by mitigating the risks of tainted and counterfeit products.[54] The OTTF intends to provide an accreditation program that will allow information and communication providers, equipment manufacturers, and those vendors that supply software or hardware components to the providers and manufacturers, to become accredited if they meet the standard's requirements and conformance criteria. Officials from DOD said that although it is unknown whether industry will adopt this standard and what the associated costs will be to maintain and use it, developing such process-based certifications along with product

---

[51] Academics and equipment manufacturers we spoke with also told us about a set of supply-chain-security best practices being developed by the Internet Security Alliance (ISA)—a multi-sector trade association whose mission is to motivate enhanced security of cyber systems. According to an ISA official, ISA has drafted a set of voluntary best practices that were developed through recommendations from industry and government. The document provides electronics manufacturers with a set of security measures for all stages of the production of electronics products that when implemented, will make it more difficult to insert malicious firmware or defective components into electronics products, such as limiting the personnel with access to design facilities to those who genuinely need to be there and using two or three factor authentication (e.g., photo radio-frequency identification and fingerprint) for employees.

[52] Officials from the Open Group told us their standards are consistent with the Office of Management and Budget Circular No. A-119, which establishes policies on federal use and development of voluntary consensus standards and conformity assessment activities.

[53] The Open Group, *Open Trusted Technology Provider Standard (O-TTPS)™ Version 1.0, Mitigating Maliciously Tainted and Counterfeit Products* (April 2013).

[54] Information and communication providers—including network providers and equipment manufacturers, government organizations, and third-party labs—participated in the OTTF's effort to establish this voluntary standard.

	certifications, such as the Common Criteria, may prove beneficial in covering more of the global IT supply chain.[55]
Communications Sector Coordinating Council (CSCC)	In accordance with Homeland Security Presidential Directive 7, the CSCC is an industry-led group that represents the viewpoints from the U.S. communications sector and facilitates coordination between industry and the federal government on improving physical and cyber security of the communications critical infrastructure.[56],[57] Representatives from the CSCC told us that the CSCC began meeting with the federal government in March 2011 to discuss supply chain security, which led to the creation of a CSCC working group to facilitate dialogue, planning, and coordination among the government and industry on supply chain risk management. This group's objectives include enhancing the government's understanding of industry's current risk management practices, the government's sharing of supply chain threat information,

[55] The Permanent Select Committee on Intelligence report cited the earlier stated concern that evaluation programs, such as the Common Criteria, that rate companies based on their processes do not address the threats because the evaluation does not include testing for vulnerabilities in the equipment. This concern could apply to the OTTF's standard because it also is based on certifying vendors' processes and not on evaluations of the equipment's integrity.

[56] Federal policy established 18 critical infrastructure sectors that are critical to the nation's security, economy, and public health and safety. The National Infrastructure Protection Plan (NIPP) presents the government's coordinated approach that will be used to establish priorities, goals, and requirements for critical infrastructure and key resources protection. The plan specifies key initiatives, milestones, and metrics to achieve the Nation's critical infrastructure and key-resources-protection mission. The NIPP also describes a partnership model as the primary means of coordinating government and private sector efforts in this area. For each sector, the model requires formation of government coordinating councils and encourages the formation of sector coordinating councils. Sector coordinating councils are self-organized, self-run, and self-governed entities comprised of critical infrastructure owners and operators that serve as the principals for sector policy coordination and planning. DHS is the sector-specific agency assigned to the communications sector that according to the NIPP, is to work with its private sector counterparts to understand and mitigate cyber risk. Department of Homeland Security, *National Infrastructure Protection Plan, Partnering to Enhance Protection and Resiliency* (2009).

[57] In February 2013, the White House released *Presidential Policy Directive 21* (PPD 21), which requires DHS to update the NIPP. PPD 21 specifically stated that the update to the NIPP "should consider sector dependencies on energy and communications systems." The White House, *Presidential Policy Directive 21* (Washington, D.C.: February 2013). According to DHS officials, following the release of the revised NIPP in late 2013, an updated communications sector-specific plan will be released, and it will address supply chain security of communications networks.

and identifying and sharing best practices for supply chain risk management. The working group is scheduled to conclude its work in December 2013.[58]

Federal Government Has Begun Efforts to Address the Risks of Using Foreign-Manufactured Equipment

The White House released an Executive Order in February 2013 that is likely to have an impact on communications supply chain security. We identified other federal efforts, such as the Interim Telecommunications Sector Risk Management Task Force, that could impact communications supply chain security, but the results of those efforts are considered sensitive, so we do not include them here.

Executive Order on Cybersecurity for Critical Infrastructure

An Executive Order released in February 2013 calls for NIST to develop a framework to reduce cyber risks to critical infrastructure and for DHS and others to spearhead increased information sharing between the federal government and owners and operators of critical infrastructure including communications networks.[59] As discussed below, federal officials told us that supply chain security may be included in these efforts, but the extent has yet to be determined.

Cybersecurity Framework

The Executive Order instructs NIST to develop a cybersecurity framework (framework) to reduce cyber risks to critical infrastructure using an open public review and comment process. This framework would provide technology-neutral guidance to critical infrastructure's owners and operators. In February 2013, NIST published a request for information (RFI) in which NIST stated it is conducting a comprehensive review to

---

[58]While the working group is currently set to end in December 2013, it may be extended beyond that date if necessary. According to one member, the working group had met twice as of December 2012.

[59]Exec. Order No. 13,636. As previously mentioned, the Executive Order seeks to improve the protection of critical infrastructure.

develop the framework and is seeking stakeholder input.[60] According to NIST officials, the extent to which supply chain security of commercial communications networks will be incorporated into the framework is largely dependent on the input it receives from stakeholders. The officials added that while it is reasonable to assume that they may receive comments about supply chain security, which crosses critical infrastructure sectors, it is possible they may not receive comments specific to the use of foreign-manufactured equipment in commercial communication networks.

In adopting the preliminary framework, the Executive Order requires agencies with responsibility for regulating the security of critical infrastructure[61] to provide a report—in consultation with national security staff, DHS, and the Office of Management and Budget—which states whether the agencies have clear authority to establish requirements based on the framework and whether any additional authorities are necessary. DHS officials stated that without seeing the context of the report, they could not say whether it would identify authorities specifically related to the supply chain security of commercial communications networks and the conditions under which those authorities could be used.

Information Sharing

The Executive Order also calls for the federal government to increase information sharing with owners and operators of critical infrastructure, including communications networks, information sharing that could

---

[60]National Institute of Standards and Technology. *Developing a Framework To Improve Critical Infrastructure Cybersecurity* (February 2013), accessed March 4, 2013, https://federalregister.gov/a/2013-04413. The RFI seeks comments on several topics including current risk management practices; use of frameworks, standards, guidelines, and best practices; the applicability of existing publications, including those of other governments; and specific industry practices. NIST has invited responses from owners and operators of critical infrastructure; federal agencies; state and local governments; standard-setting organizations; and other stakeholders.

[61]FCC, to the extent permitted by law, is to exercise its authority and expertise to partner with DHS and the Department of State, as well as other Federal departments and agencies on (1) identifying and prioritizing communications infrastructure; (2) identifying communications sector vulnerabilities and working with industry and other stakeholders to address those vulnerabilities; and (3) working with stakeholders, including industry, and engaging foreign governments and international organizations to increase the security and resilience of critical infrastructure within the communications sector and facilitating the development and implementation of best practices. The White House, *Presidential Policy Directive 21* (Washington, D.C.: February 2013).

include sharing of supply chain-related threats.[62] The order directs DHS to share unclassified cyber threat information and expand a voluntary information-sharing program that provides classified cyber threat information to critical infrastructure owners and operators with government security clearances. DHS officials told us that they foresee that this information sharing could encompass threats originating in the supply chain.

Other Approaches to Supply Chain Risk Management

Risk Management Approaches from Selected Countries

Australian Reform Proposal

The Australian government is considering a reform proposal to establish a risk-based regulatory framework to better manage national security challenges to Australia's telecommunications infrastructure.[63] The Attorney-General, in consultation with industry, has created a proposal that addresses supply chain risks by introducing a universal obligation on

---

[62]Federal agencies have multiple cyber-threat information-sharing mechanisms in partnership with the private sector, though these do not always address supply chain concerns. The mechanisms include the National Coordinating Center/Communications Sector Information Sharing and Analysis Center, Network Security Information Exchange, the National Security Telecommunications Advisory Committee, Cybersecurity Information Sharing and Collaboration Program, National Infrastructure Coordinating Center, and the United States Computer Emergency Readiness Team.

[63]For the purposes of security, Australia's telecommunication industry is regulated primarily under two pieces of legislation—the Australian Telecommunications Act (1997) administered by the Minister for Broadband, Communications and the Digital Economy and the Australian Telecommunications (Interception and Access) Act (1979) (TIA Act), administered by the Attorney-General. The TIA Act does not specifically address supply chain risks, hardware and software vulnerabilities or security risks to the confidentiality, integrity and availability of telecommunications infrastructure. See Australian Government, Attorney-General's Department, *Equipping Australia Against Emerging and Evolving Threats* (July 2012).

carriers and carriage service providers[64] to protect their networks and facilities from unauthorized access or interference. Specifically, the proposal requires carriers and carriage service providers to be able to demonstrate competent supervision and effective controls over their networks. The government would also have the authority to use enforcement measures to address noncompliance, as described in table 1.[65]

Table 1: Key Security Requirements of Australia's 2012 Reform Proposal

Key security requirements	Description
Competent supervision	Carriers and carriage service providers would be required to maintain 1) oversight (either in-house or through a trusted third party) of their network operations and the location of data; 2) awareness of, and authority over, parties with access to network infrastructure; and 3) a reasonable ability to detect security breaches or compromises.
Effective control	Carriers and carriage service providers would be required to maintain direct authority or contractual arrangements which ensure that their infrastructure and the information held on it are protected from unauthorized access. This could include arrangements to terminate contracts for security breaches and remove information and network systems where unauthorized access to a network has occurred.
Demonstration of compliance	Carriers and carriage service providers would be required to demonstrate compliance through steps such as compliance assessments and audits.
Enforcement measures/penalties for noncompliance	Government enforcement options include the authority to direct carriers and carriage service providers to undertake targeted mitigation of security risks, including modifications to infrastructure, audits, and ongoing monitoring, with costs covered by the carriers and carriage service providers; and financial penalties. The Attorney-General would also retain the power to order carriers and carriage service providers to stop service for the most serious security breaches.

Source: GAO analysis of Australian reform proposal.

Under this framework, the government would provide guidance to inform carriers and carriage service providers how they can maintain competent supervision and effective control over their networks and educate carriers and carriage service providers on national security risks. The approach

---

[64] Australia defines a "carrier" as an owner of a telecommunications facility that is used to supply carriage services to the public. It defines a "carriage service provider" as an entity that supplies a carriage service to the public using a telecommunications facility.

[65] See Australian Government, Attorney-General's Department, *Equipping Australia against Emerging and Evolving Threats* (July 2012).

would require amendments to telecommunications statutes, such as the Telecommunications Act and other relevant laws.[66]

India's Licensing Amendments

India enacted a new approach in 2011 through its operating licenses for telecommunications service providers.[67] India's Department of Telecommunications (DoT) is responsible for granting operating licenses to India's telecommunications service providers. In May 2011, DoT issued amendments to its operating licenses that included new or revised requirements for providers and equipment vendors to improve the security of India's telecommunications network infrastructure.[68] Under the amendments, telecommunications service providers are to be completely responsible for security of their networks, including the supply chain of their hardware and software. Key security requirements are described in table 2.

---

[66] Australian Government, Attorney-General's Department, *Equipping Australia against Emerging and Evolving Threats* (July 2012).

[67] In addition to the licensing approach, in February 2012, India also adopted a Preferential Market Access designed, in part, to address unspecified security concerns of the Indian government. The policy provides preference to electronic products manufactured in India in government procurements. According to the Office of the United States Trade Representative (USTR), the policy also anticipates requiring private firms to ensure that their purchases of "electronic products which have security implications" are domestically manufactured. USTR officials told us the federal government and industry, joined by other governments and foreign industry associations have raised concerns with the government of India regarding the scope and substance of this approach.

[68] Government of India, Department of Telecommunications, Letter to All Unified Access Service Licensees, No. 10- 15/2011-AS.III/(21), (May 31, 2011) (amending license clause 41.6A). USTR and others have reported that India's previous amendments to telecommunications service licenses included several controversial requirements for foreign vendors, including the forced transfer of technology to Indian companies, the escrowing of source code and other high-level and detailed designs, and assurances against malware and spyware during the entire use of the equipment. According to USTR, in response to concerns raised by industry and trading partners, including the United States, India suspended implementation of the license amendments while it consulted interested parties to better evaluate the extent to which those requirements in fact addressed India's security challenges.

Table 2: Key Security Requirements of India's 2011 Licensing Amendments

Key security requirements	Description
Organizational security policies	Providers must have an organizational policy on security and security management of their networks and must audit their own networks or contract with a network-security audit and certification agency to provide a network audit at least once a year.
Local testing requirements	Beginning April 1, 2013, all network equipment must be tested and certified to relevant Indian or international security standards in Indian labs.
Recordkeeping	Telecommunications service providers must keep a record of the supply chain of their hardware and software.
Inspection provisions	Vendors must permit the providers, DoT, or its designee to inspect the hardware, software, design, development, manufacturing facility and supply chain and subject all software to a security/threat check at any time.
Enforcement measures/penalties	DoT can issue financial penalties for inadvertent security breaches or acts of intentional omissions, such as a deliberate vulnerability left in equipment. In addition, DoT may cancel the license of the provider and blacklist the vendor that supplied the hardware or software that caused the security breach.

Source: GAO analysis of India's May 2011 Licensing Amendments.

United Kingdom's Security Requirements and Cybersecurity Evaluation Centre

The United Kingdom (UK) enacted new security and resilience requirements for network and service providers in 2011 through revisions to its Communications Act of 2003.[69] The UK's Office of Communications (Ofcom), the independent regulator and competition authority for the UK communications industries, is responsible for enforcing the requirements. According to Ofcom officials, these requirements address supply chain risks by focusing on the ability of the network and service providers to manage the overall security of their infrastructure and maintain network availability. Ofcom officials told us they are still developing their overall approach to enforcing the requirements, which are described in table 3.

---

[69]See Section 105A-D of the UK Communications Act of 2003. The UK government introduced the new security and resilience requirements, which were effective as of May 2011, to implement changes required by revisions to the regulatory framework set by the European Commission. This framework applies to all transmission networks and services used for electronic communications in European Member States. See, Ofcom, *Ofcom Guidance on Security Requirements in the Revised Communications Act 2003* (February 2012).

Table 3: Key Security Requirements for UK Network and Service Providers Enacted in 2011

Key security requirements	Description
Risk management	Network and service providers must take appropriate measures to manage risks to the security of the networks including management of general security risks; protecting end users; protecting interconnections; and maintaining network availability.
Incident reporting	Network and service providers must notify Ofcom of security breaches or reductions in availability that have a significant impact on the network or service.
Demonstration of compliance	Providers must demonstrate that a basic range of security measures have been taken. This could include compliance with security standards, such as ISO 27000 and ND1643.[a]
Enforcement measures/penalties	Ofcom could issue binding instruction to direct a provider on the steps that must be taken to improve the security of their network. For serious requirements breaches, Ofcom can impose financial penalties.

Source: GAO analysis of UK security requirements.

[a] ND 1643 is a minimum security standard for network interconnection developed by Network Interoperability Consultative Committee, a technical forum for the UK communications sector that develops interoperability standards for public communications networks and services in the UK.

A Chinese network equipment manufacturer voluntarily partnered with the UK government to establish a Cybersecurity Evaluation Centre to test its equipment for use in UK networks. According to officials from Ofcom and the Chinese manufacturer, the facility was created in part to address national security concerns related to using equipment from a vendor that did not have an established relationship with the UK government or UK network providers. The Chinese manufacturer provides the facility with the design and source code for all equipment, which is then tested for vulnerabilities by staff with UK security clearances. According to officials from Ofcom and representatives from the Chinese manufacturer, network providers cannot use the equipment until it has been approved through the testing process. In addition, the UK government requires all software patches be tested using the same process before they are installed on the equipment by the network providers. According to officials from the Chinese manufacturer, this voluntary approach helped increase trust with its customers. However, in November 2012, the chairman of the UK parliament's intelligence and security committee confirmed to us that the committee is reviewing the commercial relationship between the Chinese manufacturer and a British telecommunications provider and the Chinese manufacturer's overall presence in the UK's critical national infrastructure.[70]

Expanding Use of the U.S. Process for Reviewing Foreign Acquisitions

The U.S. government's Committee on Foreign Investment in the United States (CFIUS) conducts reviews to determine whether certain transactions that could result in foreign control of U.S. businesses pose risks to U.S. national security.[71] Industry representatives from the U.S. Communications Sector Coordinating Council told us the council and participating federal entities are discussing whether a voluntary notification process similar to CFIUS should be used for network provider purchases of foreign-manufactured equipment. In addition, the House Intelligence Permanent Select Committee report recommended that legislative proposals seeking to expand CFIUS to include purchasing agreements should receive thorough consideration by relevant congressional committees.[72]

CFIUS follows a process established by statutes and regulations for examining certain transactions that could result in foreign control of U.S. businesses. Parties generally submit voluntary notices of transactions to CFIUS, but CFIUS also has the authority to initiate reviews unilaterally.[73] Pursuant to the Foreign Investment and National Security Act of 2007,[74]

---

[70]Representatives from the UK parliament's intelligence and security committee declined to provide additional details about the inquiry.

[71]The members of CFIUS include the heads of the Departments of Treasury, Justice, Homeland Security, Commerce, Defense, State, and Energy, and Offices of the U.S. Trade Representative and Science and Technology Policy. The following offices also observe and, as appropriate, participate in CFIUS's activities: Office of Management and Budget, Council of Economic Advisors, National Security Council, National Economic Council, and Homeland Security Council. The Director of National Intelligence and the Secretary of Labor are non-voting, ex-officio members of CFIUS with roles as defined by statute and regulation.

[72]Permanent Select Committee on Intelligence, U.S. House of Representatives, *Investigative Report on the U.S. National Security Issues Posed by Chinese Telecommunications Companies Huawei and ZTE* (Washington, D.C.: Oct. 8, 2012).

[73]31 C.F.R. §§ 800.401 (procedures for notice), 800.402 (contents of voluntary notice), See also, Department of Treasury, "Committee on Foreign Investment in the United States Process Overview," accessed January 8, 2013, http://www.treasury.gov/resource-center/international/foreign-investment/Pages/cfius-overview.aspx.

[74]Pub. L. 110–49, 121 Stat. 246 (2007), amending the Defense Production Act of 1950, § 721, Act of Sept. 8, 1950, ch. 932, 64 Stat. 798, codified at 50 App. U.S.C. § 2170.

CFIUS must complete a review of a covered transaction[75] within 30 days.[76] In certain circumstances, following the review, CFIUS may initiate an investigation that may last up to 45 additional days.[77,78] If CFIUS finds that the covered transaction presents national security risks and that other provisions of law do not provide adequate authority to address the risks, then CFIUS may enter into an agreement with, or impose conditions on, the parties to mitigate such risks. If the national security risks cannot be resolved and the parties do not choose to abandon the transaction, CFIUS may refer the case to the President, who can choose whether to suspend or prohibit the transaction.[79,80] As shown in table 4, presidential decisions are rare. Table 4 also shows the number of CFIUS covered transactions, withdrawals, and other outcomes from calendar years 2009 to 2011.

---

[75]The term "covered transaction" means any merger, acquisition, or takeover that is proposed or pending after August 23, 1988, by or with any foreign person, which could result in control of a U.S. business by a foreign person. 50 App. U.S.C. § 2170(a)(3); 31 C.F.R. §§ 800.207 and 800.224.

[76]50 App. U.S.C. § 2170(b)(1)(E); 31 C.F.R. § 800.502 (beginning of thirty-day review). See also, Department of Treasury, "Committee on Foreign Investment in the United States Process Overview," accessed January 8, 2013, http://www.treasury.gov/resource-center/international/foreign-investment/Pages/cfius-overview.aspx.

[77]31 C.F.R. §§ 800.503 (determination of whether to undertake an investigation), 800.504 (determination not to undertake an investigation), 800.505 (commencement of investigation), 800.506 (completion or termination of investigation and report to the President). See also, Department of Treasury, "Committee on Foreign Investment in the United States Process Overview," accessed January 8, 2013, http://www.treasury.gov/resource-center/international/foreign-investment/Pages/cfius-overview.aspx.

[78]Parties to a transaction may request withdrawal of their notice at any time during the review or investigation stages. CFIUS must approve the requests and may include conditions on the parties, such as requirements that they keep CFIUS informed of the status of the transaction or that they re-file the transaction at a later time. See 31 C.F.R. § 800.507 (withdrawal of notice). CFIUS tracks withdrawn transactions. See Department of Treasury, "Committee on Foreign Investment in the United States Process Overview," accessed January 8, 2013, http://www.treasury.gov/resource-center/international/foreign-investment/Pages/cfius-overview.aspx.

[79]See Department of Treasury, "Committee on Foreign Investment in the United States Process Overview," accessed January 8, 2013, http://www.treasury.gov/resource-center/international/foreign-investment/Pages/cfius-overview.aspx.

[80]If CFIUS finds that the transaction in a notice does not present any national security risks or that other provisions of law provide adequate and appropriate authority to address the risks, then CFIUS will advise the parties in writing that CFIUS has concluded all action for the transaction.

Table 4: Committee on Foreign Investment in the U.S.'s Covered Transactions, Withdrawals, and Presidential Decisions, Calendar Years 2009 to 2011

Year	Number of covered transactions	Number of reviews concluded	Number of covered transactions withdrawn during review	Number of investigations concluded	Number of covered transactions withdrawn during investigations	Presidential decisions
2009	65	35	5	23	2	0
2010	93	52	6	29	6	0
2011	111	70	1	35	5	0
Total	269	157	12	87	13	0

Source: GAO analysis of Department of Treasury data.

Discussions between the Communications Sector Coordinating Council and participating federal entities on adapting a CFIUS-type voluntary notification process for use on equipment purchases are ongoing, and it is not clear how the proposal will develop, if at all.[81] The council is trying to understand the threats the government is concerned about and whether these could be best addressed by a CFIUS- type process or some other means. According to some members of the council, options range from a simple notification process, wherein network providers notify the federal government of proposed equipment purchases, to a complete review and approval process of the proposed transactions, including the aforementioned 30-day review and 45-day investigation periods.[82]

---

[81] Similarly, in its discussion paper describing its reform proposal, the Australian government noted that it initially proposed using a notification obligation for procurements in place of the requirement to provide information to the government on request. The Australian government also indicated that industry expressed a preference for an approach that avoids the need for government approval of network architecture at a technical or engineering level and instead focuses on the security outcome, leaving industry to choose the most effective way to achieve it. See, Australian Government, Attorney-General's Department, *Equipping Australia Against Emerging and Evolving Threats* (July 2012).

[82] As previously mentioned, the Interim Telecommunications Sector Risk Management Task Force is also considering a voluntary transactional review process, where network providers notify the government when they make equipment purchases or significant changes to their networks.

Potential Issues Related to Use of These Approaches

While these approaches are intended to improve supply chain security of communications networks,[83] they may also create the potential for trade barriers, additional costs, and constraints on competition. Additionally, there are other issues specific to the approach of expanding the CFIUS process to include foreign equipment purchases. We identified these issues based on interviews with foreign government officials and U.S. industry stakeholders, and our review of foreign proposals and other documentation. While the issues we identified provide a range of considerations that U.S. federal agencies would need to take into account if they chose to implement these approaches, they do not represent an exhaustive list.[84]

Trade Barriers and Disputes

Some of the approaches may create a trade barrier or cause trade disputes. The Office of the United States Trade Representative (USTR) has reported that standards-related measures that are non-transparent, discriminatory, or otherwise unwarranted can act as significant barriers to U.S. trade.[85] USTR has reported concerns regarding some of India's licensing requirements for telecommunications service providers including the following:

- the requirement for telecommunications equipment vendors to test all equipment in labs in India;
- the requirement to allow the service provider and government agencies to inspect a vendor's manufacturing facilities and supply chain and perform security checks for the duration of the contract to supply the equipment; and
- the imposition of strict liability and possible blacklisting of a vendor for taking inadequate precautionary security measures, without the right to appeal and other due process guarantees.[86]

---

[83]This is not an exhaustive list of all approaches. See appendix I for more detail on selection criteria.

[84]See appendix I for more detail on selection criteria for the factors.

[85]Office of the United States Trade Representative, *2011 Report on Technical Barriers to Trade* (Washington, D.C.: 2012).

[86]Office of the United States Trade Representative, *2012 Section 1377 Review On Compliance with Telecommunications Trade Agreement,* (Washington, D.C.: 2012). USTR officials and other industry stakeholders are working with the Indian government to help ensure that U.S. can participate in the Indian market, while respecting the security concerns of its government.

These requirements may result in trade-distorting conditions by making it more expensive and burdensome for foreign equipment manufacturers to do business in India. According to USTR, it is too early to evaluate whether the proposed reforms in Australia, new requirements and voluntary Cybersecurity Evaluation Centre in the UK, and an extension of CFIUS to equipment purchases would create trade barriers or cause trade disputes. Three U.S.-based equipment manufacturers told us that extending CFIUS to equipment purchases could cause other countries to implement similar policies, which may result in barriers to entry in other countries and trade disputes.[87]

Costs

All of the approaches may increase costs to industry and the federal government. The Australian and UK governments recognize that changes to the regulatory framework would include a cost to industry, which may increase prices for consumers.[88] Representatives from the Chinese equipment manufacturer stated that although voluntarily setting up the Cybersecurity Evaluation Centre was expensive, it was the cost of doing business in the UK. Similarly, one telecommunications industry group reported that India's 2011 License Amendments would increase compliance costs for Indian telecommunications services providers.[89] The majority (6 of 8) of equipment manufacturers we spoke with told us that any proposal to extend CFIUS to equipment purchases would increase costs for network providers, equipment manufacturers, and ultimately consumers. In addition, it is likely that the responsible federal agencies will also incur additional administrative costs in implementing any supply chain risk management requirements.

Impact on Business Decisions and Competition

All of the approaches may have an impact on the business decisions of network providers and equipment manufacturers and competition within the industry. The Australian government is aware that its proposed framework could have effects on the industry, and it is trying to anticipate

---

[87]Some of the federal entities we interviewed were not willing to discuss questions about extending CFIUS to network provider purchases of foreign-manufactured equipment.

[88]Australian Government, Attorney-General's Department, *Equipping Australia Against Emerging and Evolving Threats,* July 2012 and Ofcom, *Ofcom Guidance on Security Requirements in the Revised Communications Act 2003* (February 2012).

[89]Kent Bressie and Madeleine Findley, "Coping with India's New Telecom Equipment Security Requirements and Indigenous Innovation Policies," *Submarine Telecoms Forum*, no. 62 (2012).

these effects and explore how they might be mitigated. It is also seeking input from industry and government stakeholders on any potentially broader effects on competition in the telecommunications market and on consumers.[90] Similarly, a telecommunications industry group reported the Indian requirements complicate the relationship between telecommunications service providers and their equipment vendors, creating concerns about access to intellectual property and giving each an incentive to shift the risk of enforcement onto the other (though the current requirements still place the principal obligations on the licensees).[91] Representatives from a U.S.-based equipment manufacturer told us that extending the CFIUS process to equipment purchases could potentially lead to vendors being excluded from the U.S. market without appeal rights; this would result in limited competition and therefore potentially higher prices for consumers. Similarly, four network providers and one think tank also told us that extending CFIUS to equipment purchases would limit competition and raise costs.

Appropriate Transactions to Include in Procurement Reviews

The appropriate universe of equipment supply contracts that would be subject to review would need to be defined if the CFIUS process were extended to cover these transactions. There were 269 notices of transactions covered by the CFIUS process from 2009 through 2011. By comparison, four network providers and two equipment manufacturers we spoke with noted that network providers conduct thousands of transactions a year and expressed concerns about their being subject to a CFIUS-type process. Specifically, the two manufacturers said it would be difficult for CFIUS members to oversee all of these transactions in a timely fashion, adding expense to the procurement process for network providers and equipment manufacturers that could be passed on to consumers. In addition, CFIUS member agencies may incur significant administrative costs if asked to review thousands of procurement transactions per year.

---

[90] Australian Government, Attorney-General's Department, *Equipping Australia against Emerging and Evolving Threats* (July 2012).

[91] Kent Bressie and Madeleine Findley, "Coping with India's New Telecom Equipment Security Requirements and Indigenous Innovation Policies," *Submarine Telecoms Forum*, no. 62 (2012).

Chairman Walden, Ranking Member Eshoo, and Members of the Subcommittee, this completes my prepared statement. I would be pleased to respond to any questions that you may have at this time.

Contact and Acknowledgments

If you or your staff members have any questions about this testimony, please contact me at (202) 512-2834 or goldsteinm@gao.gov. Contact points for our Offices of Congressional Relations and Public Affairs may be found on the last page of this statement. GAO staff who made key contributions to this testimony are listed in appendix V.

Appendix I: Scope and Methodology

We focused our review on the core networks that constitute the backbone of the nation's communications system and the equipment—such as routers, switches and evolved packet cores—that transport traffic over these networks. We also focused on the wireline, wireless, and cable access networks used to connect end users to the core wireline networks. We did not address broadcast or satellite networks because they are responsible for a smaller volume of traffic than other networks.

To obtain information on all of our objectives we conducted a literature review and semi-structured interviews with or obtained written comments from academics, industry analysts, and research institutions; federal entities; domestic and foreign equipment manufacturers; industry and trade groups; network providers; and security and software audit firms as shown in table 5.

Table 5: Individuals and Organizations Selected for Interviews

Stakeholder category	Name
Academics, industry analysts, and research institutions	Center for Strategic and International Studies
	Dr. Diganta Das, Research Staff Center for Advanced Life Cycle Engineering University of Maryland, College Park
	Dr. Sandor Boyson Research Professor & Co-Director Supply Chain Management Center Robert H. Smith School of Business University of Maryland, College Park
	Gartner, Inc.
Federal entities	Department of Commerce
	Department of Defense
	Department of Homeland Security
	Department of Justice, FBI
	Department of State
	Department of Treasury
	Federal Communications Commission
	General Services Administration
	Office of the U.S. Trade Representative
	National Security Agency
	U.S.-China Economic and Security Commission
	U.S. International Trade Commission
	White House national security staff

Appendix I: Scope and Methodology

Stakeholder category	Name
Domestic and foreign equipment manufacturers	Alcatel-Lucent SA
	Cisco
	Fujitsu[a]
	Huawei Technologies
	Intel[b]
	Juniper Networks
	L.M. Ericsson
	Tellabs
	ZTE Solutions
Industry and trade groups	Association of Public Safety Communications Officials
	Communications Sector Coordinating Council
	CTIA – The Wireless Association
	Internet Security Alliance
	The Open Group
	Software Assurance Forum for Excellence in Code
	Telecommunications Industry Association
Network providers	AT&T
	Century Link
	Clearwire
	Cox Communications
	Frontier
	MetroPCS
	Sprint/Nextel
	T-Mobile
	Verizon
	Windstream
Security and software audit firm[c]	Electronic Warfare Associates (EWA)
Foreign countries	International and Regulatory Development Group
	United Kingdom
	Attorney-General's Department
	Australian Government

Source: GAO.

[a] At the time of our interview, Fujitsu no longer manufactured routers and switches, but provided aggregation and transport networking equipment.

[b] Intel makes microprocessors that are used in routers and switches.

[c] We attempted to contact McAfee as a security and software audit firm; however, it referred us to Intel representatives since McAfee is a subsidiary of Intel.

Appendix I: Scope and Methodology

We selected the stakeholders based on relevant published literature, our previous work, stakeholders' recognition and affiliation with a segment of the communications industry (domestic and foreign equipment manufacturers, industry and trade groups, network providers and so forth), and recommendations from the stakeholders interviewed. We also spoke with federal entities that have a role in addressing cybersecurity, international trade, and the Committee on Foreign Investment in the U.S. (CFIUS).

To describe how communications network providers and equipment manufacturers help ensure the security of foreign-manufactured equipment that is used in commercial communications networks, we interviewed network providers; domestic and foreign equipment manufacturers (routers, switches, and evolved packet cores); and industry and trade groups. Information we collected included specific industry practices, such as the use of security provisions in contracts; the extent to which domestic and international standards help guide their practices; and any industry-wide efforts addressing supply chain security. We focused this work on the five wireless and five wireline network providers with the highest revenue, the eight manufacturers of routers and switches with the highest market shares in the U.S. market, and two cable network providers. To identify the top five U.S. wireless providers by subscribers, we used data from the Department of Defense and verified the subscribership data against investor relations reports from the providers. To identify the top five U.S. wireline providers by subscribers, we used publicly available rankings and verified the subscriber data against investor relations reports from the providers. We selected the top eight manufacturers of routers and switches based on 2010 U.S. market share. We selected two of the top three U.S. cable companies based on 2011 subscriber data.[1]

To identify how the federal government is addressing the potential risks of foreign-manufactured equipment that is used in commercial communications networks, we asked federal agencies to identify statutes and regulations to identify potential sources of the federal government's legal and regulatory authority over how communications network providers ensure the security of their U.S. commercial networks. Additionally, we interviewed and collected documentation from 13 federal

---

[1] One of the cable companies did not respond to our request for an interview.

Appendix I: Scope and Methodology

entities to identify related federal efforts, such as interagency information sharing initiatives and those with the private sector.

To determine other approaches, including those of foreign countries, for addressing the potential risks of using foreign-manufactured equipment in commercial communications networks, we conducted a literature review and interviewed stakeholders. We identified other approaches from governmental entities in Australia, India, and the United Kingdom (UK) that address supply chain risks for commercial communications networks.[2] These countries were chosen to show the variation in how foreign governments are approaching supply chain risk management. We also considered criteria such as the availability of public information on the approach to allow for a detailed review and language considerations. While the results of the data collected from these three countries may not encompass all possible approaches, it provided important insights into the approaches that some countries are using to address supply chain risks for commercial communications networks.

We reviewed documents and interviewed officials from governmental entities in Australia, India, and the UK to describe the approaches and issues that could arise from using these approaches.[3] We identified these issues based on interviews with foreign government officials and U.S. industry stakeholders, and our review of foreign proposals and other documentation. The issues identified provide a range of considerations, but is not an exhaustive list of all issues that could be considered.

We also assessed the potential for using the CFIUS review process for purchases of foreign-manufactured equipment because a voluntary notification process similar to CFIUS is being discussed by government and industry stakeholders. We reviewed the Foreign Investment and National Security Act of 2007, related regulations, and CFIUS's annual reports to Congress to describe the CFIUS process. We reviewed CFIUS's transaction data to describe the number of covered transactions, investigations, and Presidential decisions made from calendar years 2009 to 2011 to provide context. Additionally, we interviewed officials from federal agencies and industry stakeholders on how the commercial

---

[2]We attempted to include Canada in our review, but there was limited public information on its approach and Canadian officials did not respond to our request for an interview.

[3]Indian officials did not respond to our request for an interview.

Appendix I: Scope and Methodology

communications market in the United States may be affected if any of the identified approaches are used when U.S. communications companies purchase equipment manufactured in foreign countries. We conducted data reliability testing to determine that any data used are suitable for our purposes.

We conducted this performance audit from December 2011 to May 2013 in accordance with generally accepted government auditing standards. Those standards require that we plan and perform the audit to obtain sufficient, appropriate evidence to provide a reasonable basis for our findings and conclusions based on our audit objectives. We believe that the evidence obtained provides a reasonable basis for our findings and conclusions based on our audit objectives.

Appendix II: Examples of Threats to the Information Technology Supply Chain

Supply chain threats are present at various phases of the life cycle of communications network equipment. Each of the key threats presented in table 6 could create an unacceptable risk to a communications network.

Table 6: Threats to the Information Technology Supply Chain

Threat	Description and Adverse impact
Installation of hardware or software containing malicious logic	*Malicious logic* is hardware, firmware, or software that is intentionally included or inserted in a network for a harmful purpose. Malicious logic can cause significant damage by allowing attackers to take control of entire systems and thereby read, modify, or delete sensitive information; disrupt operations; launch attacks against other organizations' systems; or destroy systems.[a]
Installation of counterfeit hardware or software	*Counterfeit information technology* is hardware or software that contains nongenuine component parts or code. The Defense Department's Information Assurance Technology Analysis Center has reported that counterfeit information technology threatens the integrity, trustworthiness, and reliability of information systems for several reasons, including the facts that counterfeiting presents an opportunity for the counterfeiter to insert malicious logic or backdoors[b] into the replicas or copies that would be far more difficult in more secure manufacturing facilities.[c]
Failure or disruption in the production or distribution of critical products	Disruptions can be caused by labor or political disputes and natural causes (e.g., earthquakes, fires, floods, or hurricanes). Failure or disruption in the production or distribution of a critical product could affect the availability of equipment that is used to support the communication networks.
Reliance on a malicious or unqualified service provider for the performance of technical services	Contractors and other service providers may, by virtue of their position, have access to network hardware and software. As we have previously reported, service providers could attempt to use their access to obtain sensitive information, commit fraud, disrupt operations, or launch attacks against other computer systems and networks.[d]
Installation of hardware or software that contains unintentional vulnerabilities	*Unintentional vulnerabilities* are hardware, software, or firmware that are included or inserted in a network and that inadvertently present opportunities for compromise. The vulnerabilities identified could allow remote attackers to, among other things, cause a denial of service. A "denial of service" is a method of attack from a single source that denies system access to legitimate users by overwhelming the targeted computer with messages and blocking legitimate traffic. It can prevent a system from being able to exchange data with other systems or use the Internet.

Source: GAO analysis of unclassified government and nongovernmental data.

Note: NIST defines "information technology" as any equipment or interconnected system or subsystem of equipment that is used in the automatic acquisition, storage, manipulation, management, movement, control, display, switching, interchange, transmission, or reception of data or information. This includes, among other things, computers, software, firmware, and services (including support services).

[a] GAO, *Cyber Analysis and Warning: DHS Faces Challenges in Establishing a Comprehensive National Capability*, GAO-08-588 (Washington, D.C.: July 31, 2008).

[b] Backdoor is a general term for a malicious program that can potentially give an intruder remote access to an infected computer. At a minimum, most backdoors allow an attacker to perform a certain set of actions on a system, such as transferring files or acquiring passwords.

[c] Information Assurance Technology Analysis Center, Security Risk Management for the Off-the-Shelf (OTS) Information and Communications Technology (ICT) Supply Chain An Information Assurance Technology Analysis Center (IATAC) State-of-the-Art Report, DO 380 (Herndon, Va.: August 2010).

[d] GAO, *Information Security: IRS Needs to Enhance Internal Control over Financial Reporting and Taxpayer Data*, GAO-11-308 (Washington, D.C.: Mar. 15, 2011).

Appendix III: Examples of Supply Chain Vulnerabilities

Threat actors can introduce the threats described in appendix II by exploiting vulnerabilities at multiple points in the global supply chain. Table 7 describes examples of the types of vulnerabilities that could be exploited.

Table 7: Examples of Supply Chain Vulnerabilities

Vulnerability	Description	Threat example
Acquisition of network equipment or parts from independent distributors, brokers, or the gray market	Purchasing from a source other than an original component manufacturer or authorized reseller may increase the risk of encountering substandard, subverted, and counterfeit products. Independent distributors purchase new parts with the intention to sell and redistribute them back into the market, without having a contractual agreement with the original component manufacturer. Brokers are a type of independent distributor that work in a just-in-time inventory environment and search the industry and locate parts for customers as requested. The "gray market" refers to the trade of parts through distribution channels that, while legal, are unofficial, unauthorized, or unintended by the original component manufacturer.	Installation of counterfeit hardware or software
Lack of adequate testing for software updates and patches	Applying untested updates and patches to network components may increase an agency's risk that an attacker could insert malicious code of its choosing into a system. For example, if a contractor fails to validate the authenticity of patches with suppliers, an attacker could write a fake patch that might allow unauthorized access to information in the network.	Installation of hardware or software containing malicious logic
Incomplete information on IT suppliers	Acquiring IT equipment, software, or services from suppliers without understanding the supplier's past performance or corporate structure may increase the risk of (1) encountering substandard, subverted, and counterfeit products, or (2) providing adversaries of the United States with access to sensitive systems or information.	Installation of hardware or software containing malicious logic Installation of hardware or software that contains unintentional vulnerabilities Installation of counterfeit hardware or software Failure or disruption in the production or distribution of critical products Reliance on a malicious or unqualified service provider for the performance of technical services

Source: GAO analysis of unclassified government and nongovernmental data.

Appendix IV: Potential Sources of Authority for Taking Action to Ensure Supply Chain Security

In examining potential sources of authority related to supply chain security, we focused on DHS, FCC, and Commerce because of their roles in critical infrastructure protection. Homeland Security Presidential Directive 7 (2003) designated DHS as the sector-specific federal agency for the telecommunications critical infrastructure sector. It required DHS to set up appropriate systems, mechanisms, and procedures to share cyber information with other federal agencies and the private sector, among others. The Communications Sector-Specific Plan of the National Infrastructure Protection Plan characterizes FCC and Commerce as partners that have relevant authority and support DHS's communications critical-infrastructure protection efforts.

Department of Homeland Security

DHS has not identified specific authorities that would permit it to take action to ensure the security of the supply chain of commercial networks. Officials from DHS's Office of General Counsel stated that the Homeland Security Act might have applicable authority,[1] although this authority is not specific to the security of the supply chain of commercial networks. DHS further stated that it cannot say what specific authority it might use if it needed to take action because it has not faced a set of circumstances related to a commercial network's supply chain security requiring action.

Federal Communications Commission

Officials from FCC's Office of General Counsel stated that FCC could regulate network providers' supply chain practices to assure that the public interest, convenience, or necessity are served if circumstances warranted.[2] Specifically, FCC could impose supply chain requirements on providers of common carrier[3] wireline and wireless voice services[4] and, in

---

[1] 6 U.S.C. §§ 121(d), 131-134, 143. See, also, *Assignment of National Security and Emergency Preparedness Communications Functions,* Exec. Order No. 13,618 77 Fed. Reg. 40,779 (2012).

[2] Under the Communications Act of June 19, 1934, ch. 652, 48 Stat. 1064, codified as amended at title 47, United States Code, the FCC has authority to regulate common carriers providing communications services. See, also, 47 U.S.C. §§ 214, 307, 309(a), 316(a).

[3] A communications common carrier, such as a telephone company, provides communications services for hire to the public. 47 U.S.C. § 153(11).

[4] Title II of the Communications Act gives the FCC authority to regulate wireline common carriers. 47 U.S.C. ch 5, subchapter II, Pt. II. Commercial mobile service providers, such as wireless phone service carriers, are also treated as common carriers under Title II of the Act, to the extent they provide common carrier services. 47 U.S.C. § 332(c). Wireless carriers are also subject to regulation as Commission licensees under Title III of the Communications Act.

Appendix IV: Potential Sources of Authority for Taking Action to Ensure Supply Chain Security

specific circumstances, information services providers,[5] using FCC's authority under the Communications Act.[6] Officials stated that FCC has not yet attempted to use these sources of authority to impose regulations specifically designed to address cybersecurity threats.

FCC officials stated that because the agency has not adopted regulations or policies related to supply chain security in commercial communications networks, reliance on these sources of authority has not been tested by legal challenges in court. According to FCC officials, legislative changes to the Communications Act to provide express recognition of the agency's authority to address such threats would reduce the risk of such challenges and may facilitate adoption of supply chain security regulation.

FCC officials added that although its current legal authority could allow FCC to act to impose supply chain requirements on network providers, it has not determined the extent to which it has authority to regulate companies that manufacture network equipment. In the past, the agency regulation of equipment manufacturers has focused on interference

---

[5]Under the Communications Act, an "information service" is defined as the offering of a capability for generating, acquiring, storing, transforming, processing, retrieving, utilizing, or making available information via telecommunications, and includes electronic publishing, but does not include any use of any such capability for the management, control, or operation of a telecommunications system or the management of a telecommunications service. 47 U.S.C. § 153(24).

[6]According to FCC officials, FCC could impose supply chain mandates on wireline carriers Under Title II of the Act. See 47 U.S.C. §§ 201, 214. Under Title III, the Commission has authority to regulate and license radio spectrum, and FCC officials told us that it could impose supply chain conditions on wireless licenses to serve the public interest, convenience, or necessity. See 47 U.S.C. §§ 307, 316. According to FCC officials, the Commission could condition new wireless licenses or modify existing wireless license to impose supply chain requirements, either individually, after allowing the licensee to protest the proposed requirements, or as a class in a rulemaking. See *Committee for Effective Cellular Rules v. FCC*, 59 F.3d 1309, 1319 (D.C. Cir. 1995), In the Matter of *Spectrum and Service Rules for Ancillary Terrestrial Components in the 1.6/2.4 GHz Big LEO Bands*, 22 FCC Rcd. 19,733, 19,743-44 (¶ 23) (2007) (citing *WBEN, Inc. v. United States*, 396 F.2d 601, 617-20 (2d Cir. 1968), *cert. denied*, 393 U.S. 914 (1968)) (examples of the Commission modifying licenses in rulemaking proceedings). With respect to information services, FCC officials told us that the Commission may regulate otherwise unregulated providers of information services, under Title I of the Communications Act, if doing so is reasonably ancillary to the effective performance of the Commission's responsibilities set out in other titles of the Communications Act. In addition, to the extent an information service provider holds any FCC licenses, the agency would have direct regulatory authority over that provider. See *In the Matter of Reporting Requirements for U.S. Providers of International Telecommunications Services* 28 FCC Rcd. 575 (2013), at ¶ 83 (exercising FCC's ancillary jurisdiction).

Appendix IV: Potential Sources of Authority for Taking Action to Ensure Supply Chain Security

management. FCC officials told us that they are actively participating in discussions within the executive branch regarding supply side issues, though which agencies should take the lead on this issue has not been determined.

Department of Commerce

Commerce officials stated that Section 232 of the Trade Expansion Act of 1962,[7] as amended, could potentially provide authority for Commerce to use when communications equipment purchases pose a potential risk to national security. According to Commerce documents, Section 232 gives Commerce statutory authority to conduct investigations to determine the effect of imports on national security. If an investigation finds that an import may threaten to impair national security, then the President may use his statutory authority to "adjust imports," by taking measures recommended by the Secretary of Commerce, including barring imports of a product. Commerce has not used, or attempted to use, this authority for any cases involving the communications sector. Commerce officials stated that they reviewed this authority in 2010 in part because a major network provider was considering purchasing foreign-manufactured communications equipment from a company that the federal government believed might pose a national security threat. Since the network provider decided not to purchase equipment from that company, Commerce did not review the potential applicability of Section 232 to that transaction.

---

[7]Pub. L. No. 87-794, 76 Stat. 872 (1962), codified as amended at 19 U.S.C. ch. 7.

Appendix V: GAO Contact and Staff Acknowledgments

GAO Contact

Mark L. Goldstein, (202) 512-2834 or goldsteinm@gao.gov

Contact and Acknowledgments

In addition to the contact named above, Heather Halliwell, Assistant Director; Derrick Collins; Swati Deo; Anne Doré; Bert Japikse; Sara Ann Moessbauer; Josh Ormond; Amy Rosewarne; and Hai Tran made key contributions to this testimony.

GAO's Mission	The Government Accountability Office, the audit, evaluation, and investigative arm of Congress, exists to support Congress in meeting its constitutional responsibilities and to help improve the performance and accountability of the federal government for the American people. GAO examines the use of public funds; evaluates federal programs and policies; and provides analyses, recommendations, and other assistance to help Congress make informed oversight, policy, and funding decisions. GAO's commitment to good government is reflected in its core values of accountability, integrity, and reliability.
Obtaining Copies of GAO Reports and Testimony	The fastest and easiest way to obtain copies of GAO documents at no cost is through GAO's website (http://www.gao.gov). Each weekday afternoon, GAO posts on its website newly released reports, testimony, and correspondence. To have GAO e-mail you a list of newly posted products, go to http://www.gao.gov and select "E-mail Updates."
Order by Phone	The price of each GAO publication reflects GAO's actual cost of production and distribution and depends on the number of pages in the publication and whether the publication is printed in color or black and white. Pricing and ordering information is posted on GAO's website, http://www.gao.gov/ordering.htm. Place orders by calling (202) 512-6000, toll free (866) 801-7077, or TDD (202) 512-2537. Orders may be paid for using American Express, Discover Card, MasterCard, Visa, check, or money order. Call for additional information.
Connect with GAO	Connect with GAO on Facebook, Flickr, Twitter, and YouTube. Subscribe to our RSS Feeds or E-mail Updates. Listen to our Podcasts. Visit GAO on the web at www.gao.gov.
To Report Fraud, Waste, and Abuse in Federal Programs	Contact: Website: http://www.gao.gov/fraudnet/fraudnet.htm E-mail: fraudnet@gao.gov Automated answering system: (800) 424-5454 or (202) 512-7470
Congressional Relations	Katherine Siggerud, Managing Director, siggerudk@gao.gov, (202) 512-4400, U.S. Government Accountability Office, 441 G Street NW, Room 7125, Washington, DC 20548
Public Affairs	Chuck Young, Managing Director, youngc1@gao.gov, (202) 512-4800 U.S. Government Accountability Office, 441 G Street NW, Room 7149 Washington, DC 20548

Please Print on Recycled Paper.

www.ingramcontent.com/pod-product-compliance
Lightning Source LLC
Chambersburg PA
CBHW081906170526
45167CB00007B/3168